Division
Bingo Book

COMPLETE BINGO GAME IN A BOOK

Written By Rebecca Stark

ISBN 978-0-87386-433-6

Educational Books 'n' Bingo

Printed in the U.S.A.

DIVISION BINGO DIRECTIONS

List of Terms

Templates for Additional Terms and Clues

2 Clues per Term

30 Unique Bingo Cards

Markers

1. **Either cut apart the book or make copies of ALL the sheets. You might want to make an extra copy of the clue sheets to use for introduction and review. Keep the sheets in an envelope for easy reuse.**

2. Cut apart the call cards with terms and clues.

3. Pass out one bingo card per student. There are enough for a class of 30.

4. Pass out markers. You may cut apart the markers included in this book or use any other small items of your choice.

5. Decide whether or not you will require the entire card to be filled. Requiring the entire card to be filled provides a better review. However, if you have a short time to fill, you may prefer to have them do the just the border or some other format. Tell the class before you begin what is required.

6. There are 50 terms. Read the list before you begin. If there are any terms that have not been covered in class, you may want to read to the students the term and clues before you begin.

7. There is a blank space in the middle of each card. You can instruct the students to use it as a free space or you can write in answers to cover terms not included. Of course, in this case you would create your own clues. (Templates provided.)

8. Shuffle the cards and place them in a pile. Two or three clues are provided for each term. If you plan to play the game with the same group more than once, you might want to choose a different clue for each game. If not, you may choose to use more than one clue.

9. Be sure to keep the cards you have used for the present game in a separate pile. When a student calls, "Bingo," he or she will have to verify that the correct answers are on his or her card AND that the markers were placed in response to the proper questions. Pull out the cards that are on the student's card keeping them in the order they were used in the game. Read each clue as it was given and ask the student to identify the correct answer from his or her card.

10. If the student has the correct answers on the card AND has shown that they were marked in response to the *correct questions,* then that student is the winner and the game is over. If the student does not have the correct answers on the card OR he or she marked the answers in response to *the wrong questions,* then the game continues until there is a proper winner.

11. If you want to play again, reshuffle the cards and begin again.

Have fun!

TERMS INCLUDED

÷ or /	32
=	35
1	36
2	40
3	42
4	44
5	45
6	48
7	49
8	50
9	54
10	55
11	56
12	60
14	63
15	64
16	72
18	81
20	DIVIDE
21	DIVIDEND
24	DIVISIBILITY
25	DIVISION
27	DIVISOR
28	QUOTIENT
30	REMAINDER

Additional Terms

Choose as many division terms as you would like and write them in the squares.
Repeat each as desired.
Cut out the squares and randomly distribute them to the class.
Instruct the students to place their square on the center space of their card.

Clues for Additional Terms

Write three clues for each of your additional terms.

_____ 1. 2. 3.	_____ 1. 2. 3.
_____ 1. 2. 3.	_____ 1. 2. 3.
_____ 1. 2. 3.	_____ 1. 2. 3.

+ − X ÷	+ − X ÷	+ − X ÷	+ − X ÷	+ − X ÷
+ − X ÷	+ − X ÷	+ − X ÷	+ − X ÷	+ − X ÷
+ − X ÷	+ − X ÷	+ − X ÷	+ − X ÷	+ − X ÷
+ − X ÷	+ − X ÷	+ − X ÷	+ − X ÷	+ − X ÷
+ − X ÷	+ − X ÷	+ − X ÷	+ − X ÷	+ − X ÷
+ − X ÷	+ − X ÷	+ − X ÷	+ − X ÷	+ − X ÷
+ − X ÷	+ − X ÷	+ − X ÷	+ − X ÷	+ − X ÷

÷ or /	=
1. This sign means "divided by."	1. This sign means "equals."
2. Fill in the blank. 9 ___ 3 = 3.	2. Whatever is left of this sign has the same value as what is to the right of this sign.
3. Fill in the blank. 15 ___ 3 = 5.	3. Fill in the sign: 24 ÷ 6 ___ 4.

1

1. $4 \div 4 =$ ___
2. $8 \div 8 =$ ___
3. $17 \div 17 =$ ___

2

1. $8 \div 4 =$ ___
2. $16 \div 8 =$ ___
3. $24 \div 12 =$ ___

3

1. $9 \div 3 =$ ___
2. $18 \div 6 =$ ___
3. $24 \div 8 =$ ___

4

1. $12 \div 3 =$ ___
2. $16 \div 4 =$ ___
3. $36 \div 9 =$ ___

5

1. $25 \div 5 =$ ___
2. $30 \div 6 =$ ___
3. $40 \div 8 =$ ___

6

1. $48 \div 8 =$ ___
2. $36 \div 6 =$ ___
3. $54 \div 9 =$ ___

7

1. $49 \div 7 =$ ___
2. $21 \div 3 =$ ___
3. $42 \div 6 =$ ___

8

1. $56 \div 7 =$ ___
2. $64 \div 8 =$ ___
3. $72 \div 9 =$ ___

Division Bingo

9 1. 81 ÷ 9 = ___ 2. 63 ÷ 7 = ___ 3. 27 ÷ 3 = ___	**10** 1. 90 ÷ 9 = ___ 2. 70 ÷ 7 = ___ 3. 30 ÷ 3 = ___
11 1. 33 ÷ 3 = ___ 2. 66 ÷ 6 = ___ 3. 44 ÷ 4= ___	**12** 1. 60 ÷ 5 = ___ 2. 48 ÷ 4 = ___ 3. 72 ÷ 6 = ___
14 1. 28 ÷ 2 = ___ 2. 14 ÷ 1 = ___ 3. ___ ÷ 2 = 7	**15** 1. 30 ÷ 2 = ___ 2. ___ ÷ 3 = 5 3. ___ ÷ 5 = 3
16 1. 32 ÷ 2 = ___ 2. 16 ÷ 1 = ___ 3. ___ ÷ 4 = 4	**18** 1. ___ ÷ 3 = 6 2. ___ ÷ 6 = 3 3. 36 ÷ 2 = ___
20 1. ___ ÷ 5 = 4 2. ___ ÷ 4= 5 3. ___ ÷ 2 = 10	**21** 1. ___ ÷ 3 = 7 2. ___ ÷ 7= 3 3. ___ ÷ 21 = 1

Division Bingo

24 1. ___ ÷ 6 = 4 2. ___ ÷ 4 = 6 3. ___ ÷ 12 = 2	**25** 1. ___ ÷ 5 = 5 2. ___ ÷ 25 = 1 3. 50 ÷ 2 = ___
27 1. ___ ÷ 3 = 9 2. ___ ÷ 9 = 3 3. 27 ÷ ___ = 1	**28** 1. ___ ÷ 7 = 4 2. ___ ÷ 4 = 7 3. ___ ÷ 14 = 2
30 1. ___ ÷ 6 = 5 2. ___ ÷ 5 = 6 3. ___ ÷ 3 = 10	**32** 1. ___ ÷ 8 = 4 2. ___ ÷ 4 = 8 3. ___ ÷ 2 = 16
35 1. ___ ÷ 7 = 5 2. ___ ÷ 5 = 7 3. ___ ÷ 1 = 35	**36** 1. ___ ÷ 9 = 4 2. ___ ÷ 4 = 9 3. ___ ÷ 6 = 6
40 1. ___ ÷ 8 = 5 2. ___ ÷ 5 = 8 3. ___ ÷ 10 = 4	**42** 1. ___ ÷ 7 = 6 2. ___ ÷ 6 = 7 3. ___ ÷ 3 = 14

Division Bingo

44	45
1. ___ ÷ 11 = 4	1. ___ ÷ 9 = 5
2. ___ ÷ 4 = 11	2. ___ ÷ 5 = 9
3. ___ ÷ 2 = 22	3. ___ ÷ 3 = 15

48	49
1. ___ ÷ 8 = 6	1. ___ ÷ 7 = 7
2. ___ ÷ 6 = 8	2. ___ ÷ 1 = 49
3. ___ ÷ 4 = 12	3. ___ ÷ 49 = 1

50	54
1. ___ ÷ 10 = 5	1. ___ ÷ 9 = 6
2. ___ ÷ 5 = 10	2. ___ ÷ 6 = 9
3. ___ ÷ 2 = 25	3. ___ ÷ 3 = 18

55	56
1. ___ ÷ 11 = 5	1. ___ ÷ 8 = 7
2. ___ ÷ 5 = 11	2. ___ ÷ 7 = 8
3. ___ ÷ 55 = 1	3. ___ ÷ 2 = 28

60	63
1. ___ ÷ 12 = 5	1. ___ ÷ 9 = 7
2. ___ ÷ 5 = 12	2. ___ ÷ 7 = 9
3. ___ ÷ 10 = 6	3. ___ ÷ 63 = 1

Division Bingo

64	72
1. ___ ÷ 8 = 8	1. ___ ÷ 8 = 9
2. ___ ÷ 4 = 16	2. ___ ÷ 9 = 8
3. ___ ÷ 16 = 4	3. ___ ÷ 2 = 36

81	**Divide**
1. ___ ÷ 9 = 81	1. To separate into parts.
2. ___ ÷ 81 = 1	2. To perform the operation of division.
3. ___ ÷ 1 = 81	3. To subject a number to the process of division.

Dividend	**Divisible**
1. It is the amount to be divided into parts.	1. Capable of being divided, especially with no remainder.
2. It is the number or amount to be divided into parts by another number.	2. A number is ___ by another if, after dividing , the remainder is zero.
3. In the problem 10 ÷ 2, 10 is the ___.	3. The number 9 is ___ by 3, but the number 8 is not.

Division	**Divisor**
1. A basic arithmetic operation. The others are addition, subtraction, & multiplication.	1. The quantity by which the dividend is to be divided.
2. This operation calls for a number to be separated into parts.	2. In the problem 10 ÷ 2, 2 is the ___.
3. Long ___ is used when the divisor is a large number; each step is shown in detail.	3. In the problem 72 ÷ 9, 9 is the ___.

Quotient	**Remainder**
1. The number that results when dividing one quantity by another.	1. The amount left over after division.
2. In the problem 48 ÷ 12 = 4, 4 is the ___.	2. In the problem 11 ÷ 5, 11 is not divisible by 5, so the answer will have a ___ of 1
3. In the problem 56 ÷ 7 = 8, 8 is the ___.	3. The answer of 16 ÷ 5 is 3 R1, or 3 with a ___ of 1.

Division Bingo

30	24	81	50	9
6	÷ or /	Divisibility	49	28
Dividend	63		36	54
Quotient	64	18	44	35
55	10	7	15	20

Division
Bingo

Quotient	Dividend	40	56	81
35	49	2	64	45
72	10		8	18
14	21	63	16	28
20	Divisibility	7	6	15

Division Bingo

Quotient	18	49	44	Dividend
10	÷ or /	3	24	27
64	Divisibility		45	=
63	72	55	14	40
15	6	7	16	81

Division Bingo

63	45	Division	6	81
42	1	24	56	Dividend
36	14		9	50
18	Divide	Divisibility	7	2
25	20	48	15	54

Division Bingo

20	9	64	2	6
42	18	3	8	÷ or /
Division	54		32	11
28	81	30	16	25
49	7	Dividend	63	36

Division Bingo

=	45	40	Division	54
44	64	25	24	Dividend
56	4		1	8
7	55	16	48	81
35	18	30	36	Divide

Division Bingo

30	45	11	32	49
35	81	10	÷ or /	42
40	50		8	1
63	11	3	Quotient	72
7	6	16	48	=

Division Bingo

36	45	5	44	1
42	Division	56	54	2
Divide	60		81	9
15	63	Quotient	25	14
Divisibility	7	48	64	35

Division Bingo

8	49	10	Divide	6
25	Division	36	64	45
27	30		÷ or /	5
4	20	55	32	11
14	16	3	Quotient	9

Division Bingo

Quotient	44	1	56	Divide
54	2	24	÷ or /	Division
60	45		50	72
55	28	25	16	27
3	35	40	20	36

Division Bingo

=	45	64	25	35
5	27	32	8	24
42	81		40	10
3	Dividend	16	6	Quotient
4	7	30	48	49

Division Bingo

49	9	27	44	8
10	Divisibility	81	48	÷ or /
30	11		54	56
7	14	Division	Quotient	42
45	5	60	4	2

Division
Bingo

4	9	=	27	54
Division	5	81	8	72
44	2		10	11
36	16	1	60	Quotient
7	28	48	30	32

Division Bingo

6	Division	64	8	4
2	30	27	÷ or /	45
25	50		40	3
28	16	60	1	=
7	56	72	35	36

Division Bingo: Card No. 14

Division Bingo

32	8	64	49	44
=	40	24	81	25
54	30		Dividend	45
7	27	5	16	4
35	14	48	Divide	10

Division Bingo

1	27	5	Divide	21
56	72	11	42	50
4	9		54	10
63	2	7	32	Quotient
25	Remainder	48	14	45

Division Bingo: Card No. 16

Division Bingo

3	Divisor	12	27	6
32	25	16	50	11
8	36		Remainder	5
20	35	Quotient	64	72
55	4	49	44	9

Division Bingo

Divide	60	2	25	56
45	3	55	54	4
8	72		12	Division
20	24	16	Quotient	40
Remainder	27	64	Divisor	=

Division Bingo

54	=	27	5	60
32	44	45	49	50
Divisor	6		÷ or /	Dividend
40	Remainder	55	14	12
81	21	35	36	48

Division Bingo

60	Divisor	44	27	÷ or /
2	10	42	55	56
9	11		63	24
20	Divisibility	15	14	Remainder
18	36	21	Quotient	12

Division Bingo

32	=	42	27	28
9	12	1	5	30
72	35		Divisor	64
55	49	Remainder	20	36
63	21	48	3	14

Division Bingo

Divide	40	12	Division	4
56	44	Dividend	5	÷ or /
2	50		30	11
Remainder	20	14	24	6
21	3	Divisor	72	42

Division Bingo

1	Divisor	49	81	48
=	60	35	32	24
40	4		15	30
72	21	Remainder	3	14
28	Divisibility	36	55	12

Division Bingo: Card No. 23

Division Bingo

1	60	6	Divisor	5
12	48	42	56	30
11	Divide		4	72
28	15	Remainder	3	9
18	63	21	44	Divisibility

Division
Bingo

63	42	Divisor	64	12
24	28	32	1	÷ or /
9	5		15	Remainder
Dividend	20	Divisibility	21	50
48	6	2	25	18

Division Bingo

12	Divisor	40	56	Divide
55	44	5	60	1
28	15		50	63
3	Division	20	21	Remainder
11	25	64	Divisibility	18

Division Bingo

40	2	Divisor	60	10
28	15	32	Remainder	÷ or /
16	Divisibility		21	63
Divide	=	42	18	24
4	50	12	Dividend	11

Division Bingo

54	60	Dividend	Divisor	1
10	12	15	56	50
Divisibility	72		11	55
Quotient	Divide	35	21	Remainder
Division	8	4	18	28

Division Bingo

12	60	Divide	32	8
28	55	42	11	Dividend
9	15		÷ or /	Divisor
10	20	Division	21	Remainder
1	5	18	=	Divisibility

Division Bingo

6	Divisor	56	8	Remainder
24	60	40	50	÷ or /
28	4		11	42
18	=	Division	21	15
20	49	Divisibility	12	Dividend

Division Bingo: Card No. 30

© Barbara M. Peller

www.ingramcontent.com/pod-product-compliance
Lightning Source LLC
Chambersburg PA
CBHW080721220326
41520CB00056B/7284